An Introduction to Geological Maps

D0473314

SECOND EDITION

J. A. G. THOMAS
B.Sc., F.G.S.

Former Head of the Geography and Geology Department
Verdin Comprehensive School, Winsford, Cheshire

London
A THOMAS MURBY PUBLICATION OF
GEORGE ALLEN & UNWIN
Boston   Sydney

PREFACE

The demand for a map book containing suitably graded exercises for use during C.S.E., 'O' and 'A' level courses continues; but changes in syllabuses have led to a number of alterations designed to make this book more useful.

The elementary treatment of strike-line maps found in the previous edition is no longer necessary, and has been removed. It is hoped to incorporate this material in a new 'A' level map book. New exercises include a number of maps based, by permission, on those of the Institute of Geological Sciences but simplified where this has been thought desirable. At the request of some teachers, ornament has been omitted on a few of these maps so that they can be coloured by hand. The study and use of I.G.S. maps in the field is the way to real understanding and enjoyment, and every school should possess and use a representative selection on various scales.

One or two problem maps which require the identification of minerals, rocks and fossils or drawings of fossils have been incorporated, and in these teachers may provide or substitute their own specimens. Sketch-sections are also required in some exercises.

While every care has been taken in preparing material, errors may have crept in. Suggestions for corrections and improvements will be gratefully received and carefully considered.

J.A.G.T.

CONTENTS

KEY TO SYMBOLS USED IN MAP BOOK

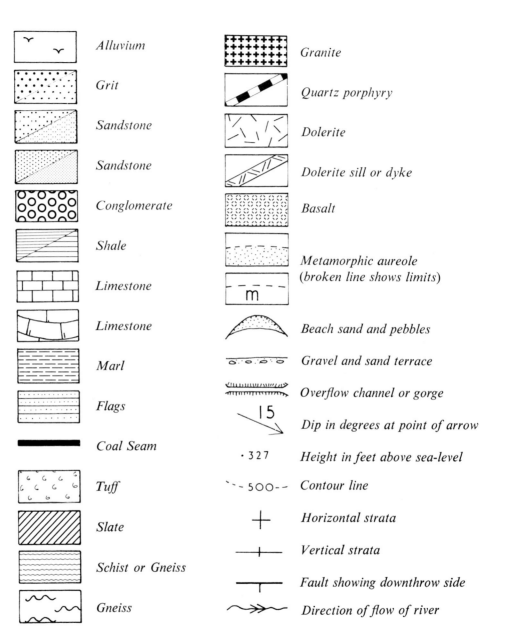

	Alluvium
	Grit
	Sandstone
	Sandstone
	Conglomerate
	Shale
	Limestone
	Limestone
	Marl
	Flags
	Coal Seam
	Tuff
	Slate
	Schist or Gneiss
	Gneiss

	Granite
	Quartz porphyry
	Dolerite
	Dolerite sill or dyke
	Basalt
	Metamorphic aureole (broken line shows limits)
	Beach sand and pebbles
	Gravel and sand terrace
	Overflow channel or gorge
15	Dip in degrees at point of arrow
· 327	Height in feet above sea-level
~ 500 ~	Contour line
	Horizontal strata
	Vertical strata
	Fault showing downthrow side
	Direction of flow of river

4

Sedimentary Rocks

HORIZONTAL BEDS

THE rocks most commonly met with on geological maps are sedimentary in origin, and most of these were laid down in almost horizontal layers as a *conformable series* on the sea floor. It may help us to think of these beds as a succession of blankets and sheets spread one upon the other. A rectangular layer cake cut out of such a pile is represented in the block diagram below (Fig. 1).

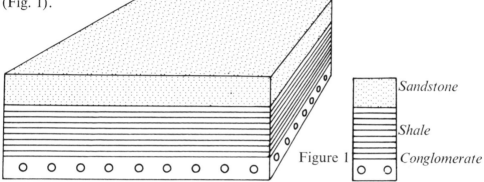

Figure 1

Sandstone

Shale

Conglomerate

Notice that the *key* or *stratigraphical index* is shown as a vertical column cut out of the block, with the beds in order of deposition, the oldest at the bottom and the youngest at the top; this is how the stratigraphical succession is shown on most geological maps, including those of the Institute of Geological Sciences. In problem maps, however, where the order of the beds has to be determined by the student, the key is given as a number of separate rectangles in any order of age, as on page 4. Reference should be made to this key.

If the block shown in Figure 1 is raised slowly until it is above the sea, its surface will be level and composed entirely of sandstone; but rivers will form, cutting valleys down through the sandstone and reaching shale (Fig. 2).

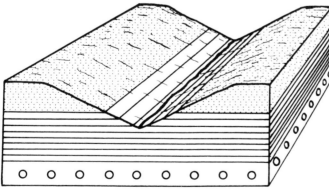

Figure 2 *Block diagram of part of river valley cut in horizontal strata.*

5

Since the bedding plane between the shale and sandstone is horizontal, it is always at the same height above sea-level, so that where it emerges on the surface it must follow a contour. In the map below (Fig. 3 *a*) it follows the 350 m contour.

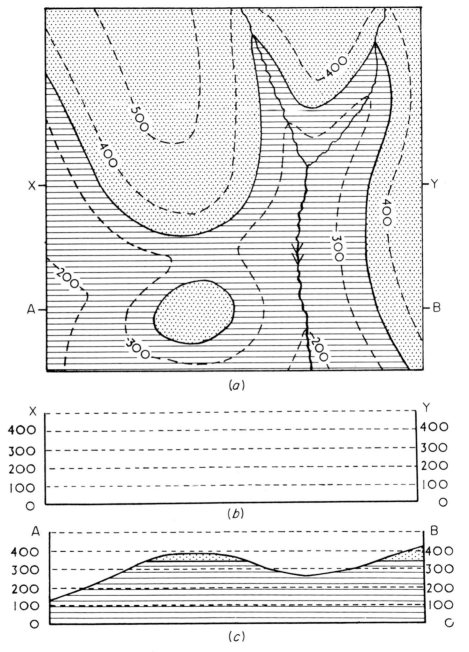

Figure 3

A *section* (or vertical cut) from *A* to *B* is shown below the map (Fig. 3 *c*).

Exercise

Complete the section from *X* to *Y* on the frame provided (Fig. 3 *b*).

On the map draw a thin pencil line from *X* to *Y* and along it place the straight edge of a slip of paper. Mark carefully on the paper the position of *X* and *Y*, the points where each contour is crossed, the river and the junction of the beds. The paper slip will then appear as in Figure 4.

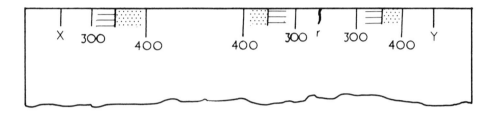

Figure 4

With the edge horizontal, slide this up and down between the vertical lines provided (Fig. 3 *b*), taking care to keep the point *X* on one line, and the point *Y* on the other. When the edge is just below the horizontal line representing a height of 300 m, transfer to the latter all the 300 m points on your slip; then slide it up almost to the next line, and repeat for the 400 m points, and so on. Remove the slip and join the points you have marked in a smooth curve, extending this sideways to the vertical lines without crossing any further horizontal lines. (Why?) Remember that the rivers you have marked should correspond with the valley bottoms. You now have a profile of the surface along the line of section. Using the slip as before, mark on this profile the junction between shale and sandstone: since the beds are horizontal, this should be at a height of 350 m and the section may now be completed.

As the section frame is exactly below and parallel to the line *XY*, it would be quicker to place a ruler along the base of the frame and, using a set-square and pencil to drop imaginary perpendiculars, transfer the required points directly. However, the method first described can be used in all cases, including those where the line of section is not parallel to the base (Figs. 13 & 64).

The isolated outcrop of sandstone shown as capping a hill in the section *AB* (Fig. 3 *c*) is completely surrounded by an older bed (shale) and is known as an *outlier*.

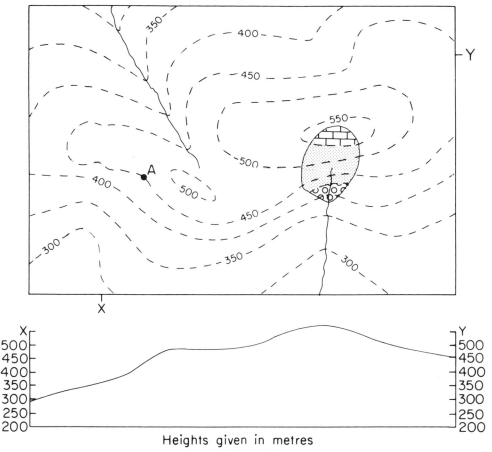

Heights given in metres

Figure 5

A borehole is put down at the point A in sandstone. 25 m below the surface the drill enters conglomerate, and passes through this formation to a depth of 150 m below the surface when it enters shale which is at least 100 m thick. A geological surveyor finds and maps the exposures shown to the east of the borehole site (see key on page 4). If all the beds are horizontal, complete the outcrops on the map, and draw a section from *X* to *Y* on the profile provided.

How to tackle this problem

First look at the exposure, and write down

   Height of base of limestone is 550 m (on contour)
   Height of base of sandstone is 425 m (between contours and in borehole)
   Height of base of conglomerate is ? m
            (work this out from the borehole log).

Now draw round the 550 m contour as the boundary of the limestone, putting the symbol inside (i.e. above). Proceed in the same way for other formations. The 425 m contour must be drawn in first.

INCLINED STRATA

Sedimentary rocks are not usually horizontal. Earth movements may tilt or fold them, and prolonged denudation has the effect of planing them, so that the bevelled edges of the strata are seen as parallel outcrops on the surface; this is shown in the block diagram (Fig. 6). Here the beds are dipping to the east, while the outcrops strike north to south. The angle of dip is the angle between the horizontal and the line of greatest slope on the bedding plane (θ in Fig 7). Note that a bed is said to outcrop whether it has a covering of soil or vegetation or not. If the rock is seen at the surface, this part of the outcrop is described as an exposure. A section taken parallel to the dip arrow is a *dip section*, and shows the true dip of the beds. A *strike section* is parallel to the strike, or perpendicular to the dip section, and shows apparently horizontal beds. Any section drawn between these directions shows an *apparent dip* which is less than the true dip.

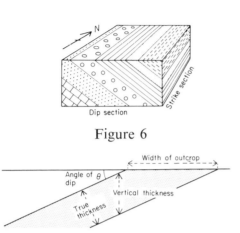

Figure 6

Figure 7

The section (Fig. 7) shows a dipping bed of sandstone. The beds above and below have been omitted for the sake of clarity.

Exercise

Draw a horizontal line (the surface) and place a ruler diagonally across it to represent the dipping bed, as in Figure 7. Draw along both edges of the ruler and mark in the width of outcrop: for angles of dip of 15°, 30° and 60° (why not 5°?) it may be seen that where the surface is horizontal the smaller the dip, the wider the outcrop. For small dips the vertical thickness and true thickness are almost the same and for dips of 10° or less the difference between them is negligible.

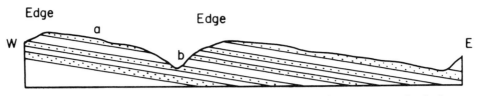

Figure 8 *The stippled beds are grit bands, alternating with shale.*

Effects of slope on outcrop

The slope of the ground also affects the width of the outcrop.

Two *edges* or *escarpments*, their dip slopes formed by resistant beds of grit, are shown in the section above (Fig. 8). Compare the width of outcrop of the grit capping the hill at *a* with its width where it outcrops on the scarp face at *b*.

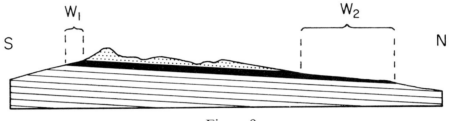

Figure 9

The bed of clay shaded black in the section above has a much narrower outcrop in the south (W_1) than in the north (W_2). Why?

Exercise

The section below (Fig. 10) shows two escarpments formed by beds of limestone in a succession of shales.

   a. Express the thickness of *one* of the limestone formations as a fraction of the thickness of the shales between them.

   b. Express the widths of outcrop of the same two formations in the same way.

   c. Explain the wide discrepancy between both fractions.

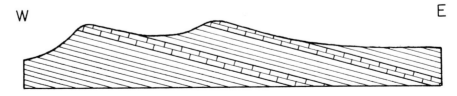

Figure 10

10

The surface of the land never slopes uniformly over great distances, and outcrops usually change direction with a change of slope.

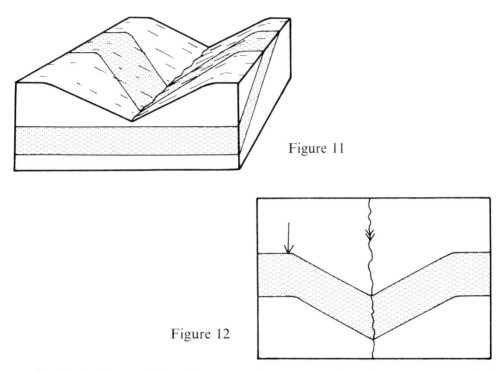

Figure 11

Figure 12

In the block diagram (Fig. 11), a river has cut a V-shaped valley across a bed dipping towards the observer, causing the outcrop to *V* (*vee*) along the valley in the direction of dip.

Figure 12 gives a bird's-eye view or map of the top, showing the outcrop of the bed, with the direction of dip marked by an arrow. The V in the outcrop can be likened to the head of a larger arrow, and the river to the shaft, so giving us a useful method of finding the direction of dip when this is not shown.

This effect can be demonstrated by opening an exercise book to form a valley, and fitting in a textbook cornerwise to represent that part of the dipping bed removed by erosion. An assistant now runs a pencil along both edges of the smaller book, front and back, to mark the outcrop of the bed. When viewed from above (as in a map) the outcrop is seen to vee in the direction of dip.

When the beds are horizontal, or are dipping at very low angles, the V of the outcrop may not always indicate the dip, but in these cases the V is generally long and narrow, for example Figure 3 *a*, where the strata are horizontal.

Exercise

Draw a geological map similar to Figure 12, but showing *two* parallel rivers flowing across a bed dipping towards the observer. State in which direction the outcrop vees on the intervening ridge.

11

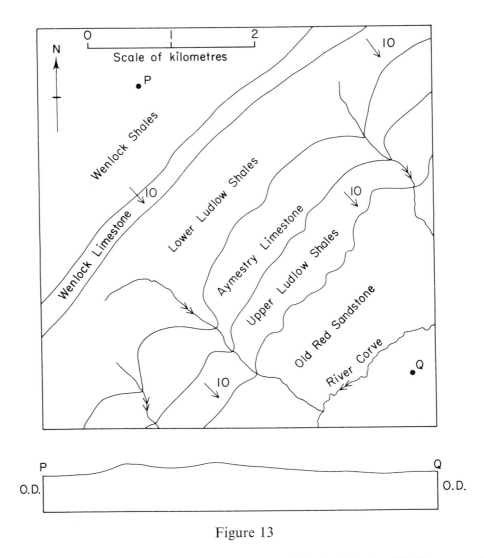

Figure 13

Figure 13 is a simplified version of part of the I.G.S. 1 : 25 000 map of Wenlock Edge. The regional dip is 10° to the S.E. so that the oldest formation, the Wenlock Shales, outcrops in the N.W. and dips under the succeeding Wenlock Limestone, and so on. Outcrops are much wider than the true thicknesses of the beds because the dips are small (see page 10). The Wenlock Limestone forms a continuous escarpment, Wenlock Edge, but the parallel escarpment of Aymestry Limestone is broken by streams flowing in steepsided valleys, with outcrops veeing along them in the direction of dip.

Exercise

Draw a section from P to Q on the profile provided.

(The boundaries may be marked in by using a slip of paper as on page 7, and the angle of dip with a protractor.)

FOLDED STRATA

Inclined strata often form one limb of a fold, two simple folds being shown in the block diagrams below.

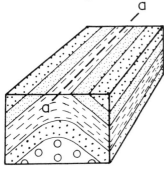

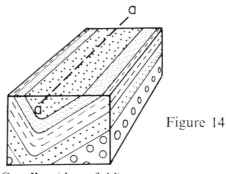

Figure 14

Anticline (upfold)

In the diagram both limbs dip at the same angle. This anticline is therefore said to be symmetrical and outcrops of the same formation have the same width on each side of the axis (aa).

Syncline (downfold)

Here the limbs dip at different angles so that outcrops of the same bed have different widths. The syncline is said to be asymmetrical.

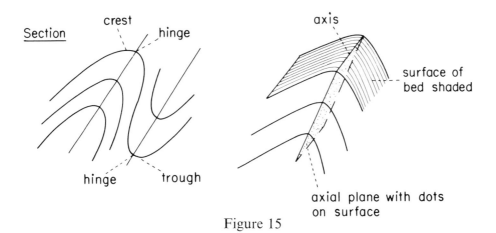

Figure 15

The *axis* of a fold is its hinge line, defined as the line joining points of maximum curvature in any single bed. Reference to Figure 15 will show that the axis does not always coincide with the line joining the highest or lowest points (*crests* and *troughs* respectively) of a folded bed.

A surface which is drawn to include axes or hinge lines in successive beds in a fold is known as the *axial plane* (Fig. 15) though in some folds it may be curved. On the ground the axis as defined above is rarely seen, but its position can generally be drawn with reasonable accuracy on maps of simply-folded formations. See also revision exercise on page 64, Figure 75.

13

Exercise

On the profile provided (Fig. 16) draw a section from *A* to *B* to show the structure. Mark on the map any fold axes present.

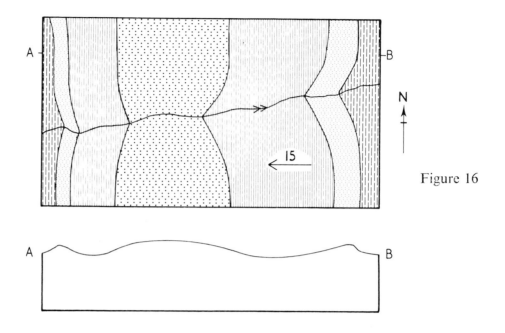

Figure 16

The dip of the shale is 15° westwards at the point of the arrow, and we assume that the dip of the adjacent formations is the same. But the shale and sandstone reappear on the other side of the grit, so a fold must be present. Since the outcrops are narrower west of the axis, the dips must be steeper there. Note also the evidence of the vees in relation to the direction and angle of dip.

Exercise

The photograph Figure 17 shows a syncline and an anticline with an almost vertical limb in common.

1. Mark in the points of maximum curvature on each bed visible in the syncline, and join them to form an axial plane. Indicate in any one bed a trough which does not lie on the axial plane.

2. Repeat this exercise for the anticline, but indicating a crest which does not lie on the axial plane.

Strictly speaking you have drawn a section or *trace* of each axial plane.

Figure 17

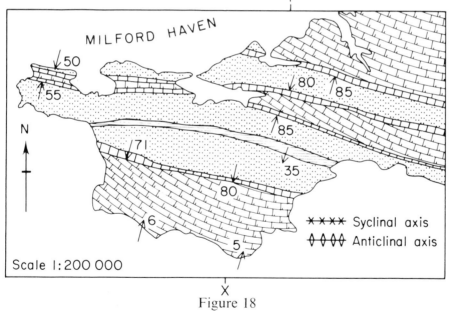

Figure 18

Exercise

Figure 18 is a simplified geological map of S.W. Pembrokeshire, now part of
Dyfed. Mark in one synclinal axis and one anticlinal axis, using the symbols
given. Draw a sketch section from X to Y. (On this scale the land surface
appears to be almost level.)

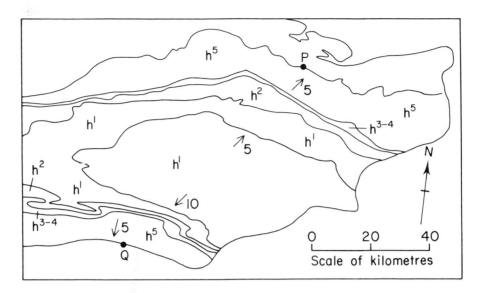

Figure 19

Exercise

Figure 19 is a simplified map of the geology of the Weald. h =Hastings Beds, h^1=Weald Clay, h^2=Lower Greensand, h^{3-4}=Upper Greensand and Gault, h^5=Chalk. Draw a sketch-section from P to Q. (On the scale of this map, relief is scarcely visible on a profile.) Map and section may be coloured with crayons.

Exercises

1. A river flowing from east to west in a V-shaped valley is crossed by a vertical bed. Draw a sketch map to show the outcrop of the bed.

2. State what structure is shown by the outcrops on this map (first mark in the dip arrows).

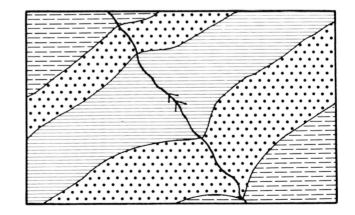

Figure 20

16

3. Using the pro-
file provided, sketch
a geological section
from *A* to *B* (method
on page 7).

a. Note the repetition
of beds as on page
14, (Fig. 16).

b. Using an exercise
book to represent
a valley, hold a
textbook *vertically*
and trace the out-
crop in pencil.
How does the val-
ley, or any in-
equality in the
surface, affect this
outcrop?

4. State in what dir-
ection the beds dip
in Figure 22.

Figure 22

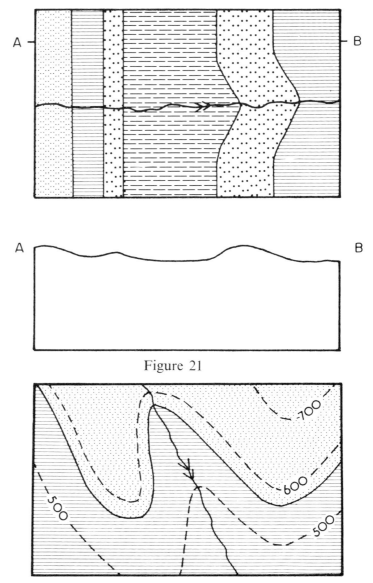

Figure 21

17

Revision

The maps in Figure 23 show identical relief and drainage. On each the outcrop of a coal seam is shown by a heavy black line.

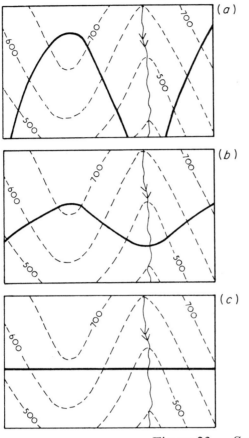

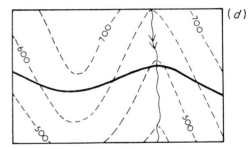

a. The seam dips *gently* towards the observer. Notice the sinuous outcrop and the long incomplete V pointing downstream.

b. The seam dips more steeply towards the observer than in *a*. The outcrop is straighter and the V shallower, but still pointing downstream.

c. The seam is vertical and the outcrop is straight.

d. The seam now dips steeply away from the observer. The V is shallow but points *up-stream*.

Figure 23   *Scale: 1 : 50 000*

Exercise

Construct a relief model of the area shown above in clay, Plasticene, or similar material. Mark in any one of the outcrops shown in (*a*) or (*b*) or (*d*). Hold the model at arm's-length and tilt until the outcrop appears as an almost straight line (because of inaccuracies in construction, it is unlikely to be exactly straight). Your eye is now on an extension of the bedding plane. The outcrop could also be obtained by cutting into the model with a long straight-edged knife held at the angle of dip.

18

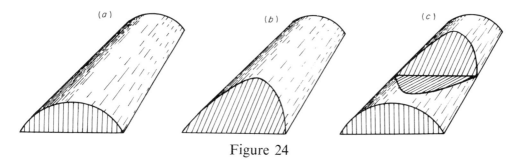

Figure 24

Figure 24 *a* represents a piece of wooden beading sawn through vertically, its cross section being the segment of a circle. In *b* the saw cut slopes down towards the observer, and in *c* a V-shaped saw cut has been made in the middle. If we make and study such a model, it will help us to understand how the outcrops of beds in an anticline are affected by slopes and by transverse valleys (valleys cutting across the axis). In a vertical cliff as in *a* the anticline can be seen in section, but only the topmost beds are seen from above; the lower beds cannot be represented on an ordinary map although they reach the surface. When an anticline outcrops on rising ground, as in *b*, the outcrop of any one bed is concave towards lower ground. In a river valley, as in *c*, there is a double concavity, and a boat-shaped outcrop is formed (see also Fig. 26). Note the V of the outcrops with the dip along the bottom of the valley. Study of the photograph of a model (Fig. 25) should make these points clearer.

Figure 25

19

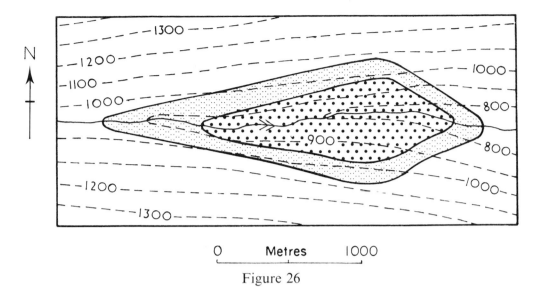

Figure 26

Study Figure 26 carefully. The river and contours show that a V-shaped valley runs from west to east. The scale indicates the steepness of the valley sides. Since the outcrops of the grit and sandstone vee along the valley, dip arrows may be inserted (see page 11). These point westwards in the western half, and eastwards in the eastern half of the map, showing that the beds are folded in an anticline. The axis runs almost north-south across the widest part of the outcrop, and should be marked in.

Exercises

1. *a*. State which limb of the anticline is dipping more steeply and give reasons for your answer.

   b. The V pointing to the west is long and narrow, so that the western limb could be horizontal (page 11). How do we know that it is, in fact, dipping westwards?

2. The section below (Fig. 27) shows the long profile from source to mouth of a river which has cut into a series of dipping beds; the broken line shows the slope of the ground on either side of the valley. Draw a geological sketch map of the valley and compare with Figure 26.

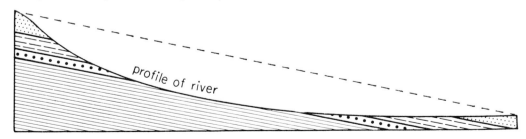

Figure 27

20

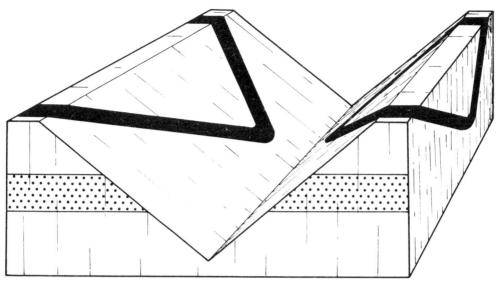

Figure 28

The block diagram above shows a syncline or downfold cut by a transverse valley. The outcrop of only one bed (in black) has been completed. Notice how it is concave towards higher ground.

Exercise

Complete the outcrops of the sandstone formation shown on the front of Figure 28.

Keep the outcrops parallel to those of the other bed.

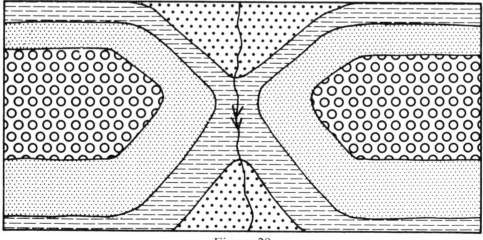

Figure 29

In Figure 29 a valley occupies approximately half the width of the map. Insert dip arrows (see page 11) and the axis of the fold. State if it is a syncline or anticline. Compare Figure 16.

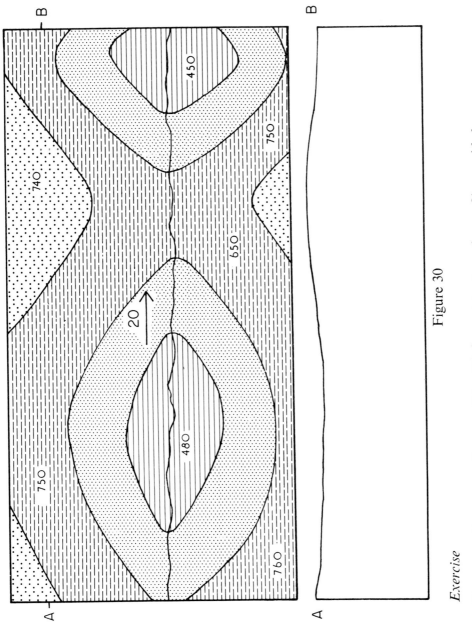

Figure 30

Exercise

Draw a section across the map (Fig. 30) from *A* to *B* on the profile provided.

Join *AB* with a pencil line, and transfer junctions of beds to the profile in the usual way (page 7). One dip is given: draw it accurately on the profile. Mark on the map other directions of dip along the valley. Are the dips at these localities about 20°, or greater, or less (page 14)? Now complete the section.

Monoclines

A monocline is shown in the block diagram below (Fig. 31). The strata on either side of the fold are horizontal, or nearly so.

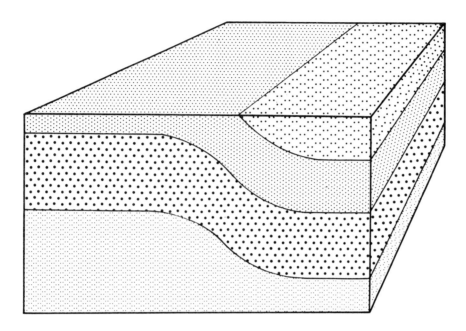

Figure 31

A simplified section through a monocline in the Isle of Wight is shown below (Fig. 32).

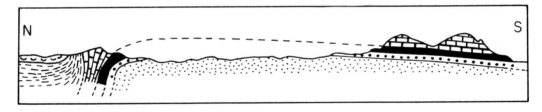

Figure 32

Notice the almost horizontal northern and southern limbs. The chalk, denoted by a symbol resembling a brick wall, has been removed from the centre by denudation: the former position of its base is indicated by a broken line.

The special 1 : 63 360 I.G.S. sheet of the Isle of Wight shows this section and the resulting outcrops in some detail and should be studied carefully.

Overfolds and isoclinal folding

When the forces producing folding are more intense, overfolds in which both limbs are more or less parallel are produced. These are shown in the block diagram below (Fig. 33).

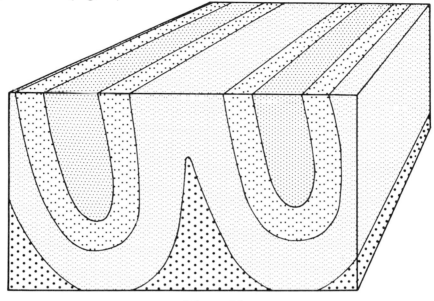

Figure 33

After denudation a geological traverse across the axes of these folds will reveal a succession of beds all dipping in the same direction at comparable angles: hence the term *isoclinal folding* given to this type of structure. It is easy to misinterpret the field evidence and to postulate a thick series of strata tilted steeply in one direction, as has happened more than once in the history of geological science. Note the repetition of beds on the top surface of the block diagram; in problem maps it may be assumed that beds represented by the same symbol are of the same age. The succession is inverted in one limb of an overfold; this may be shown either by a dip of more than 90° (for example 110° instead of 70°), or the apparent dip (70° in this case) may be given and evidence of inversion provided in other ways.

Exercises

1. In Figure 33 beds are repeated at outcrop. Draw sketch maps to show four other ways in which this may happen, two in the case of horizontal strata and two with simple folds. In each map indicate the dip.

2. Construct a solid model showing isoclinal folding, using a small cardboard box covered with white paper as your base. Insert dip arrows on the top surface (they should not all show exactly the same angle of dip).

24

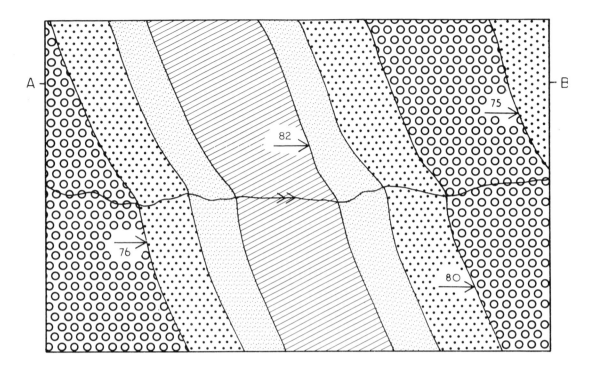

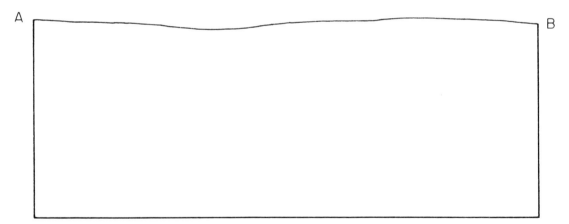

Figure 34

Exercise

Draw a geological section across the map from *A* to *B* on the profile provided.
The oldest formation shown is a conglomerate.

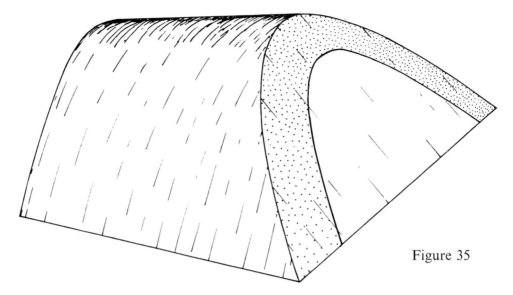

Figure 35

The block diagram above shows a bed of sandstone folded into an anticline and terminated by a slope on the right-hand side. The bed is stippled only on the slope. On a map the outcrop of the sandstone is seen to be concave down-slope (Fig. 24).

By turning the page in an anti-clockwise direction, so that the line at the base of the slope leads away from the observer, the slope appears to be horizontal, and the axis along the crest of the anticline appears to be dipping to the left-hand side. This is a plunging anticline.

In both cases—horizontal axis with sloping ground, and sloping axis with level ground—the outcrops of the sandstone exhibit the same shape. They may be distinguished by studying the slope of the ground and by the direction of dip arrows (see Exercise 1 below).

Exercises

1. Hold an old exercise book in the shape of an anticline plunging fairly steeply. On its surface draw a horizontal chalk line to represent the shape of outcrop on level ground, and on opposite sides of the axis draw arrows in the direction of dip. Now draw a map of the outcrops in a plunging anticline, showing the axis and dip arrows.

2. Repeat the above exercise for a plunging syncline.

3. If working in pairs, a thin exercise book or piece of stiff paper may be held in a double fold (anticline and adjacent syncline), and outcrops and dip arrows plotted in the same way.

26

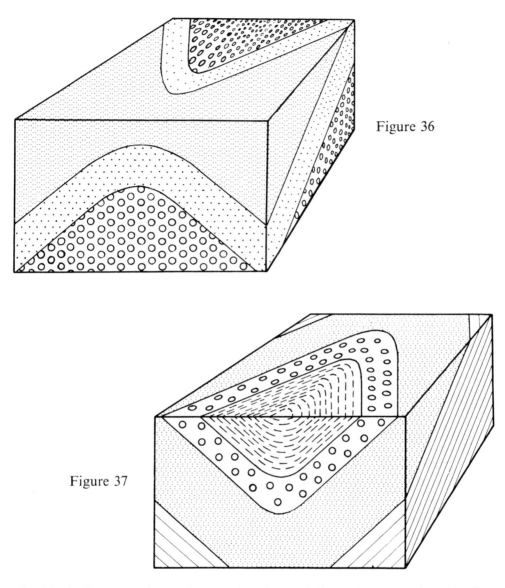

Figure 36

Figure 37

The block diagrams above show a plunging anticline (Fig. 36) and a plunging syncline (Fig. 37). Study the shape of the outcrops carefully.

Exercises

1. Draw a block diagram showing a double fold—syncline and anticline together—plunging away from the observer.

2. The map on page 28 is based on a 1 : 10 560 I.G.S. map, slightly simplified. Draw a section from *A* to *B* on the profile provided. What evidence given on the map shows that the folds are plunging?

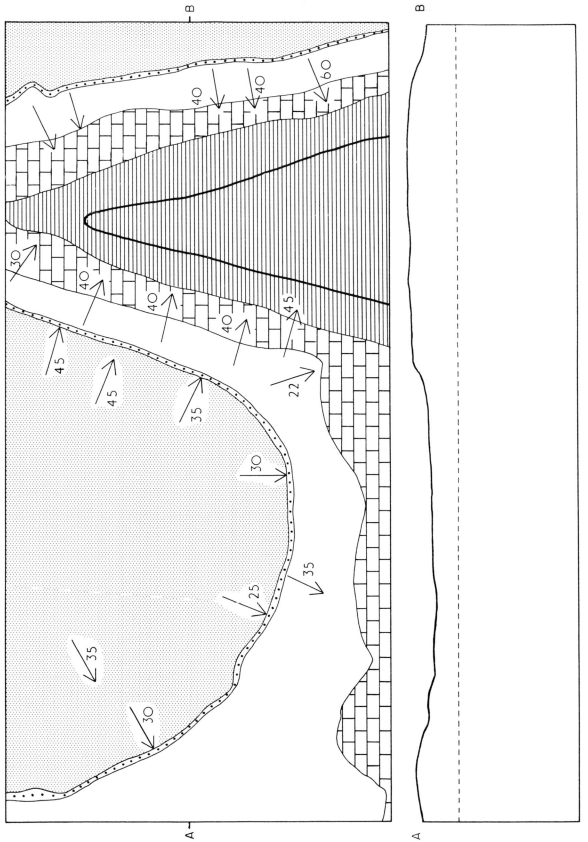

Figure 38

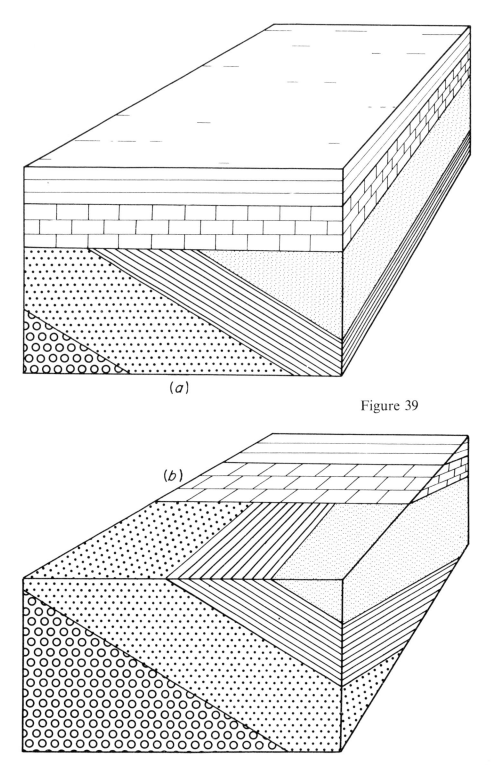

(a)

Figure 39

(b)

29

UNCONFORMITY

Figure 39 *a* shows an older conformable series of beds (conglomerate below, followed by grit, shale and sandstone) which have been tilted down to the right and planed off by erosion. On the almost level surface so formed, a newer series of limestone and clay formations has been deposited. This is known as an *unconformity*, and the newer series is said to lie *unconformably* on the older. A plane of unconformity separates the two series.

If the newer series consists of marine sedimentary rocks, they must have been uplifted since their deposition, and such uplift is usually accompanied by some degree of tilting. Figure 39 *b* shows the effect of such tilting followed by further erosion.

Notice the basal bed (limestone) of the newer series. It rests in turn on different beds of the lower series, both at depth and at outcrop. Such a relationship is known as *overstep*, and is the way in which an unconformity may be recognized on a map. Figure 40, adapted from the 1 : 63 360 I.G.S. map of Shrewsbury, shows overstep of Cambrian shales (unshaded) and Ordovician formations by the basal Silurian Grit.

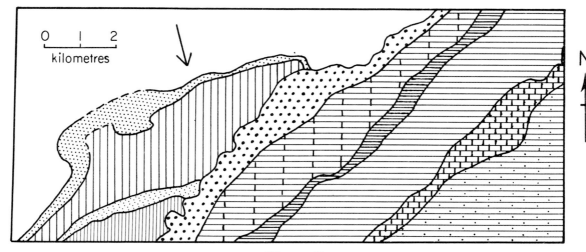

Figure 40

Exercises

1. Mark the plane of unconformity present in Fig. 40 with a heavy line, and draw an arrow to show the direction of dip in the younger series. Why are the strikes in both series almost parallel?

2. A series of strata is tilted and eroded. A younger series is laid down unconformably on this surface and is subsequently uplifted, the whole being tilted still further in the same direction as before. Draw a block diagram to show this structure after erosion has revealed the older series.

Use textbooks for the older series, exercise books for the newer series.
Photographs of unconformities will be found on pages 66 and 67.

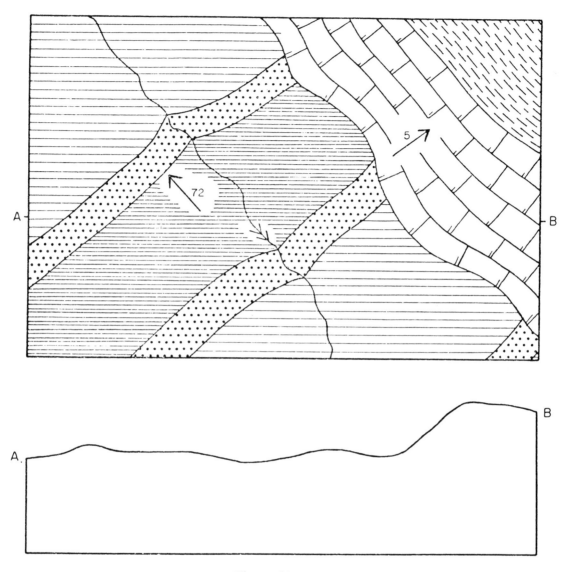

Figure 41

Exercise

Draw a section across the map from *A* to *B* on the profile provided below.

A section taken parallel to the direction of dip uses the actual dip shown in degrees; if taken at right angles, there is no dip in section. A section between the two directions has an intermediate dip; this can readily be found by calculation or construction, but is not required in elementary map work. Note also the repetition of beds. Draw the axes of the folds, to cross the line of section. Are these folds present below the unconformable cover?

31

FAULTS

A fault is a fracture in the earth's crust, the rocks on one side being displaced relative to those on the other.

The fracture is conventionally spoken of as a fault plane and is usually depicted as such on a map, though in fact it may be slightly curved. It is sometimes represented by a zone of fault breccia or broken rock (Fig. 76).

Normal faults

These are the normal or usual faults encountered in coalfields. The fault plane is generally steeply inclined, the rocks on one side having apparently moved down the slope so formed. Thus in Figure 42 the left-hand side has moved down relative to the right-hand side and is known as the *downthrow* side, the right-hand side being the *upthrow* side.

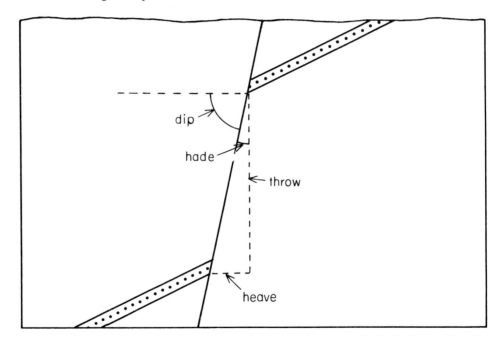

Figure 42

In the section above (Fig. 42) one bed only is shown. The vertical distance moved is known as the *throw*, the horizontal distance as the *heave*. The angle between the fault plane and the vertical is known as the *hade*, but this traditional mining term is being superseded by *dip*, used with its ordinary meaning as the angle between the fault plane and the horizontal.

Though the total throw of a fault may amount to hundreds of metres, the movement is seldom more than a few metres at a time. Weathering attacks the upthrow side more rapidly, tending to remove any inequality in height.

32

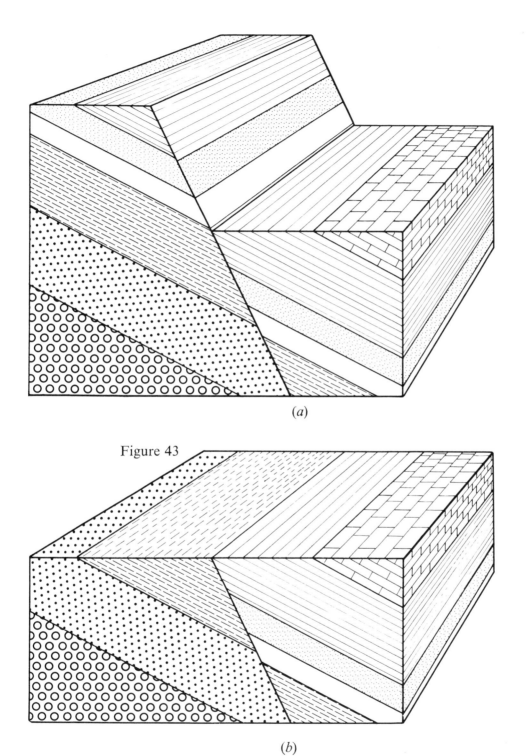

(a)

Figure 43

(b)

33

The effect of a normal fault is shown in more detail in the block diagrams on the previous page. In Figure 43 *a* the right-hand side has moved downwards relative to the left-hand side, forming a fault-scarp. Such an original feature, modified by erosion, is found only in localities where fault-movement has been comparatively recent and rapid. Figure 43 *b* shows the same block after erosion has levelled the area. Often there is no topographical indication of faulting at the surface.

Notice that the strata and the fault dip in the same direction. The effect of faulting has been to cut out two complete formations and part of the shale at outcrop: there is no surface indication that they are present.

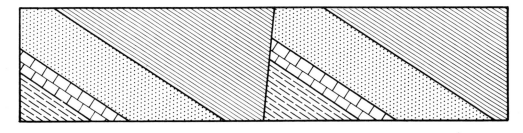

Figure 44

The section in Figure 44 also shows a normal fault dipping and downthrowing against the dip of the beds. In this case there is repetition of the outcrops; the amount of repetition depends on the magnitude of the throw and the dip of the beds involved.

Exercises

1. Using the scale of 1 cm to 100 m (1 : 10 000) find

    a. The throw of the fault in Figure 43,

    b. the *thickness* of strata cut out in Figure 43,

    c. the throw of the fault in Figure 44,

    d. the thickness of strata repeated in Figure 44.

2. Study the figures and text on pages 33 to 35, and complete the following table:

Effect of Faults on Outcrops

	Normal Fault	*Reverse Fault*
Throw with the dip.	*Some beds cut out.*	
Throw against the dip.		

Reversed faults

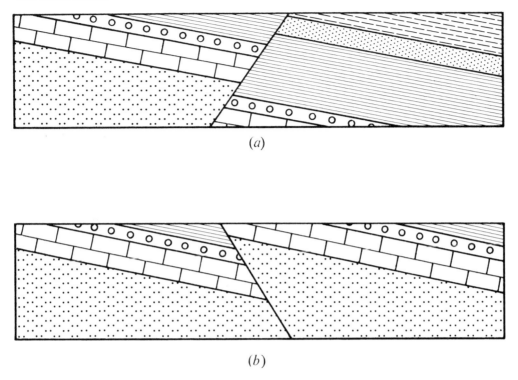

(a)

(b)

Figure 45

Sections across *reversed* or *reverse* faults are shown in Figure 45. These should be carefully studied, and Exercise 2 on the previous page completed. Note that movement is up the inclined fault plane. From previous exercises it should be apparent that a vertical fault always appears as a straight line on a map, one with a high dip is almost straight, while faults with low dips tend to be sinuous where the relief is varied. In this respect they act exactly as bedding planes in dipping strata (Fig. 23). If the angle of dip is very small, a reversed fault is called a thrust.

Exercises

1. Cover a small cardboard box with white paper, and use it as the basis of a model showing on one side the section in Figure 44. The exercise may be varied to include Figure 45 *a* or *b*.

2. From Sheet 1 of the I.G.S. 1 : 633 600 map of Britain trace the Moine Thrust (the most easterly thrust, running from Loch Carron, NG 8937, to Loch Eriboll, NC 4054), the Great Glen Fault (from Loch Linnhe to the Moray Firth), and the Highland Boundary Fault (from Loch Lomond to Stonehaven). Explain the differences you see in these traces.

Dip and strike faults

All the faults so far described have been parallel to the strike of the strata and are known as strike faults. Dip faults are more or less parallel to the direction of dip, while faults running at a marked angle to the dip and strike are often called oblique faults.

Figure 46 shows a normal dip fault with downthrow to the left, the beds dipping towards the observer.

Figure 46

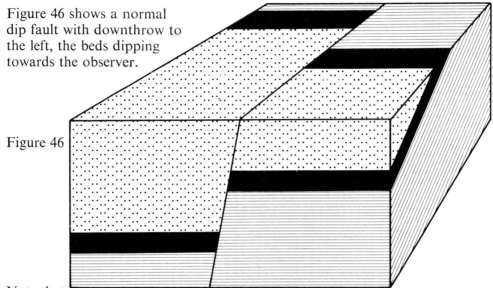

Note that

(i) the outcrops appear to be displaced sideways (lateral displacement);

(ii) if the outcrops of two formations are in contact across a fault, *the younger formation is on the downthrow side*.

Exercise

Draw a section across the map (Fig. 47) from *A* to *B* on the profile provided.

First note the unconformity. Show it on the profile dipping at *less* than 5°. Next mark in the fault; it is a straight line, so is vertical. Using the rule (ii) above, mark the downthrow side with a short straight line (see Fig. 48 for symbol). The strike of the beds it affects is parallel to the line of section, so they have no apparent dip. To find the thickness of the shale, draw a section parallel to the dip arrow (18) as in Figure 7, assuming the surface to be level.

Geological History

An older series of rocks, with grit below, followed in order by shale and sandstone, was laid down, tilted down to the south and faulted, then eroded. After subsidence limestone was deposited, then followed uplift with further slight tilting to the south-east, the limestone being removed from the western part of the area by erosion to reveal the older series.

36

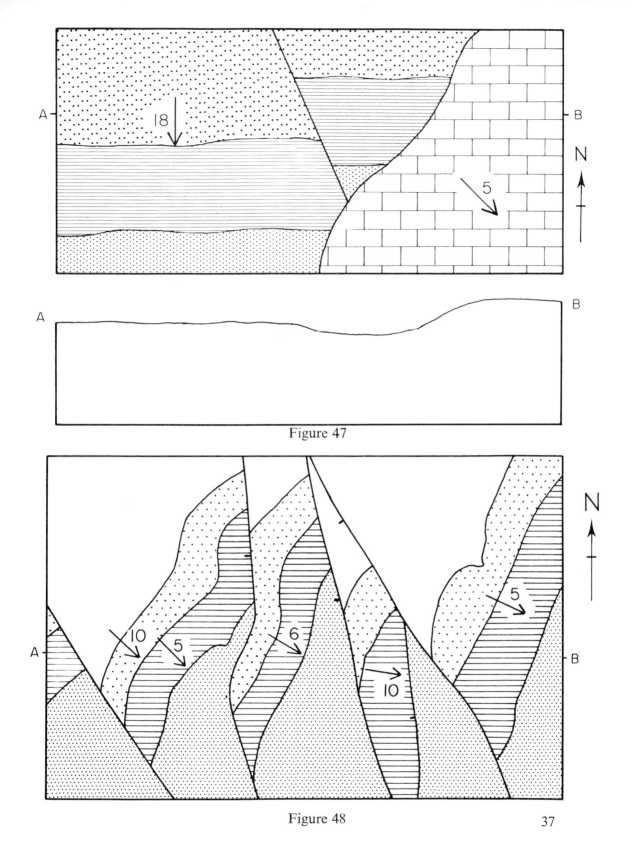

Figure 47

Figure 48

Figure 48 is based on part of a 1 : 63 360 I.G.S. map (Runcorn, Sheet 97). Notice the number of normal faults, all trending in a general NNW-SSE direction, with downthrow to the west shown by the usual symbol. They cannot be unconformities, for a formation cannot rest unconformably on itself. The dip of the beds is low, but varies slightly from exposure to exposure.

Exercises

1. In Figure 48, draw a sketch section from *A* to *B*. (No relief is indicated so assume that the surface is level.)

2. In Figure 49, draw a section on the profile provided.

   Identify and draw the plane of unconformity first. The conglomerate is puzzling; it is a local (i.e. restricted) deposit, laid down possibly in hollows by the advancing sea. Show it as a wedge in section. Next draw the fault. It is not vertical (why?), so show it dipping steeply to the downthrow side. The shale and grit are repeated, so suspect a fold. Insert dip arrows to show the unshaded beds dipping under the grit on both sides of the axis.

3. Describe the geological history of the area shown in Figure 49 (see account on page 36).

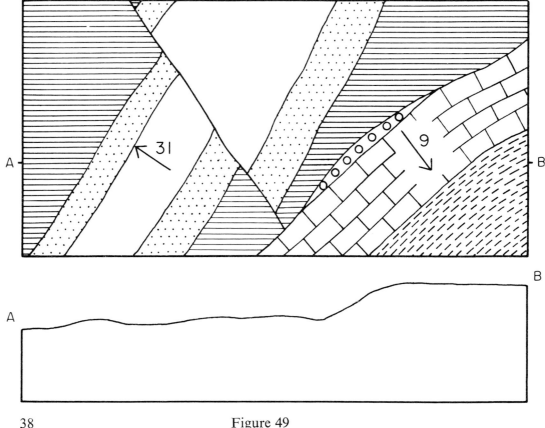

Figure 49

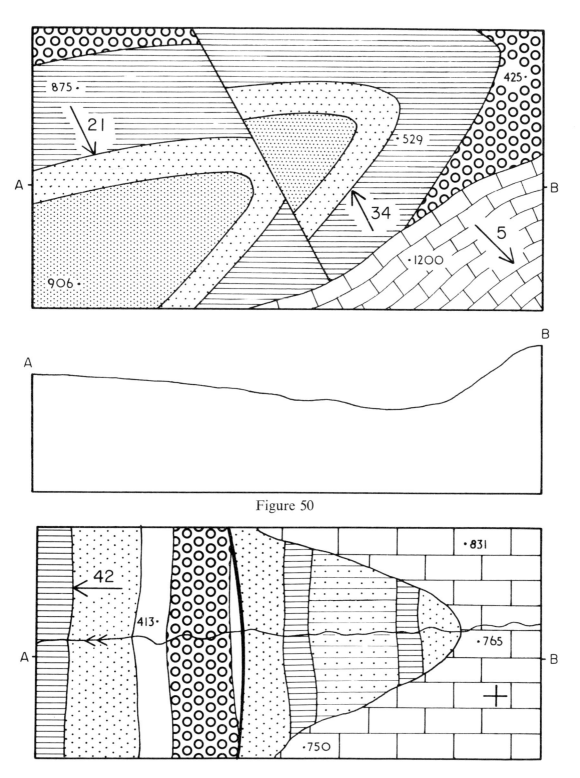

Figure 50

Figure 51

Exercises

1. Draw a section across the map from *A* to *B* on the profile provided (Fig. 50).

Is the fold plunging, or does the ground rise? (Exercise 1 and 2 on page 26). Which is the downthrow side of the fault? (Note (ii), page 36).

2. Draw a sketch section across the map (Fig. 51) from *A* to *B*.

The grit is dipping *under* the shale; insert dip arrows indicating this for all the grit-shale contacts. What other bed is repeated? The fault is shown as a thick black line.

3. Draw a section across the map from *A* to *B* on the profile provided (Fig. 52). Faults are shown by thick black lines.

Make a preliminary sketch to see which way the faults are inclined.

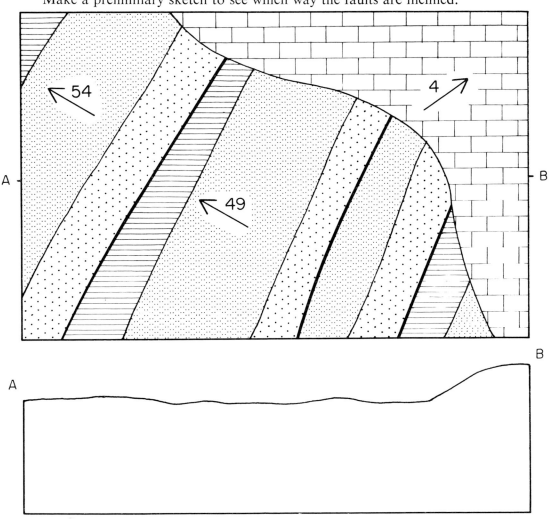

40   Figure 52

Tear faults

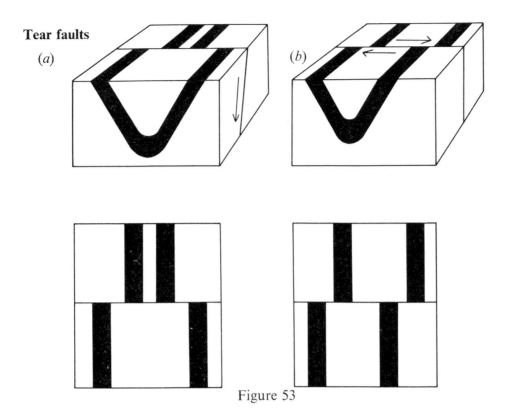

Figure 53

The block diagrams and maps above omit all but one formation for the sake of clarity.

Figure 53 *a* shows a syncline crossed by a normal fault, the arrow showing the direction of movement conventionally. It is, of course, possible that both sides have moved up or down by different amounts.

Figure 53 *b* shows a similar syncline also crossed by a fault, but the displacement of outcrops on the far side can be explained only by sideways movement. Such a fault is a *tear* (pronounced 'tare') or *wrench* fault, the Great Glen Fault of Scotland being one of the best known British examples.

Study the maps below the block diagrams and insert dip arrows. Be prepared to recognize these patterns again.

Exercises

1. Explain how you would distinguish on a map between folded strata affected by a tear fault and isoclinal folds crossed by a normal fault.

2. Draw block diagrams showing an adjacent anticline and syncline,

    a. crossed by a tear fault,

    b. crossed by a normal fault.

The geology of the Wren's Nest, near Dudley. (Reproduced by permission of the Nature Conservancy Council.)

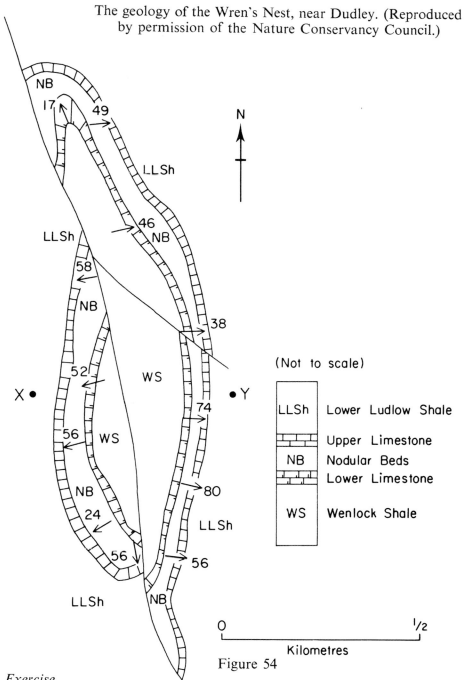

Figure 54

Exercise

Mark in the downthrow side of each fault. Draw a sketch section from X to Y, projecting beds and fault above the surface with broken lines to indicate the original structure before erosion. What name could be applied to this structure?

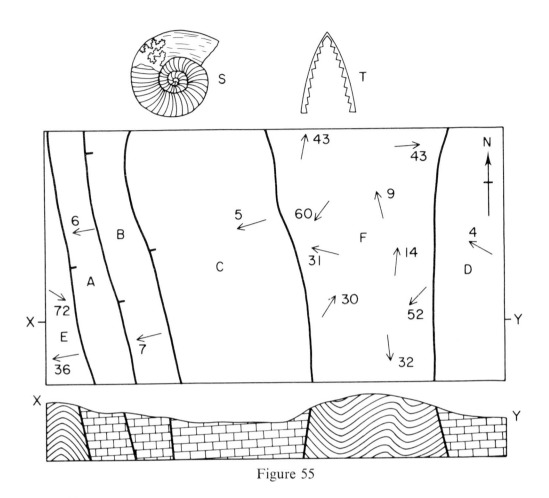

Figure 55

Problem

Limestone at A, B, C and D contains fossils of type S, while indurated shales at E and F yield fossils of type T. Suggest a geological age for (a) the limestone; (b) the shales. Indicate the downthrow side on the two normal faults where this is not shown, and complete the section.

Solution

Graptoloid graptolites are Lower Palaeozoic in age, while tuning-fork grapto-lites can be assigned to the Ordovician. The ammonoid (shown by its suture to be an ammonite) is of Mesozoic age. Both faults downthrow to the younger side, i.e. towards the limestone. The section can now be completed as shown, the complex folding in the shales being drawn conventionally.

The three most westerly faults throw down to the east in succession, and are termed *step faults*. The upthrown block of shale bounded by normal faults is a *horst*, and the downthrown block of limestone west of it a *graben*. Such structures do not always produce topographic features.

43

Revision Exercise

Draw sketch sections across each of the maps below between the points marked on the sides. In (6) assume that the limestone is horizontal.

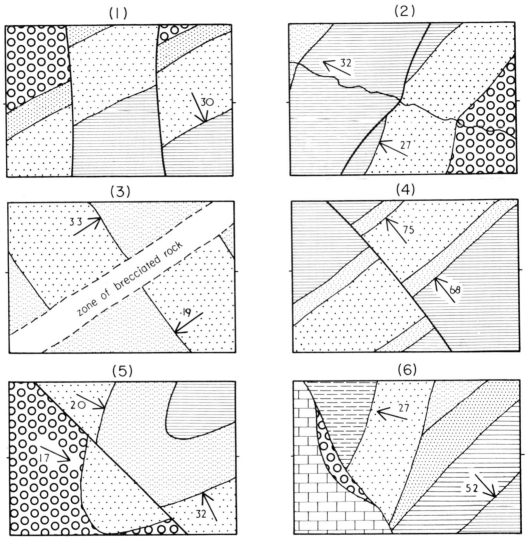

(1) (2) (3) (4) (5) (6)

Figure 56

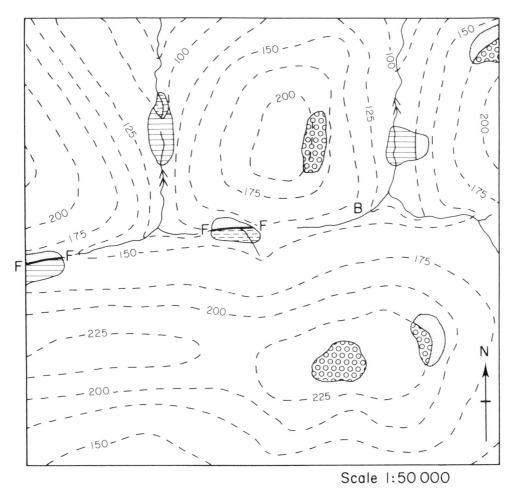

Scale 1:50 000

Figure 57

Exercises

Figure 57 shows a contoured map with exposures marked in by a geological surveyor. He found during his survey that all the rocks were horizontally bedded, and that a fault, F-F, was visible in two of the exposures. At B he could see brecciated rock in the stream bed.

1. Mark in the probable course of the fault. Give reasons why you think that the fault follows the line you have drawn.

2. Draw a stratigraphical index (page 5) to scale showing the succession of formations present. (It is easier if you concentrate on the area north of the fault). Show the height above sea level of the base of each formation where possible.

3. Complete the map by drawing the outcrops north of the fault.

4. State the throw of the fault, and mark in its downthrow side.

5. Complete the outcrops south of the fault.

45

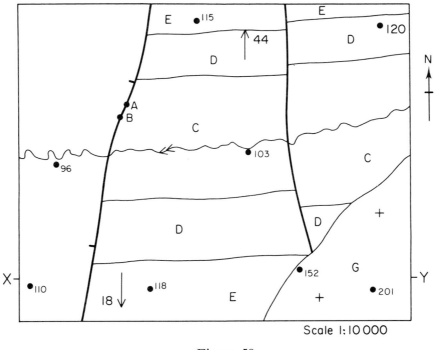

Figure 58

(Six specimens are required for this exercise. Suitable minerals for A and B include quartz, calcite, barite, fluorite, galena, blende and pyrite. Specimens C, D, E and G may be of any suitable sedimentary rock.)

Exercise

You are provided with six specimens A, B, C, D, E and G; a streak plate, a steel scratching needle, a 'copper' coin and dilute hydrochloric acid.

1. State the tests and observations you have made on specimens A and B, and their results. Identify each specimen.

2. Explain why specimens A and B are associated with a fault.

3. Describe, test and identify specimens C, D, E and G, giving reasons for your identification.

4. Study the map, Figure 58, and explain why the trace of the more easterly of the two faults stops when it reaches formation G.

5. Mark in the downthrow side of this fault.

6. Complete the outcrops west of the other fault, assuming that the throws of both faults are approximately equal. Check by using Note (ii) on page 36.

7. Draw a sketch-section from X to Y.

46

Igneous Activity

Magma finds its way to the surface along planes of weakness such as faults, joints and other fractures which are vertical or steeply inclined, consolidating as wall-like bodies known as *dykes* (Fig. 59). Since these cut across original structures such as bedding planes, they are said to be *discordant* and are younger than the rocks they intrude. If magma forces its way along bedding planes it forms *sills*. These *concordant* intrusions are folded and faulted with the sedimentary beds they intrude and behave like them at outcrop. Sometimes they transgress from one horizon to another (Fig. 59). They are younger than the rocks intruded, but older than the folds.

A lava flow is extrusive, cooling and solidifying at the surface. If submarine, it may be interbedded with sedimentary rocks of marine origin; but, if terrestrial, it may rest unconformably on older strata. Thermal metamorphism caused by sills, dykes, and lava flows is usually limited in extent, and is rarely shown on maps; for example, shales may be baked for a few inches on both sides of the intrusion. Lava flows can affect only the rocks beneath.

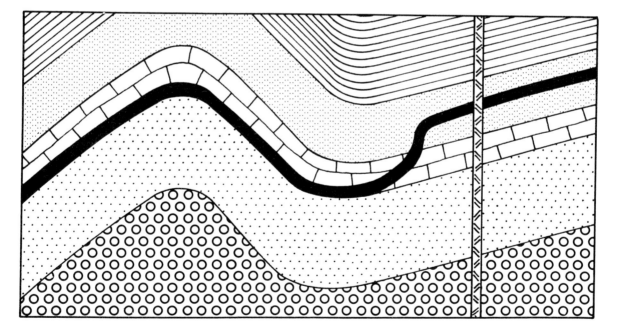

Figure 59. *Section showing folded strata with a dyke (vertical) and a sill (black) transgressing the bedding.*

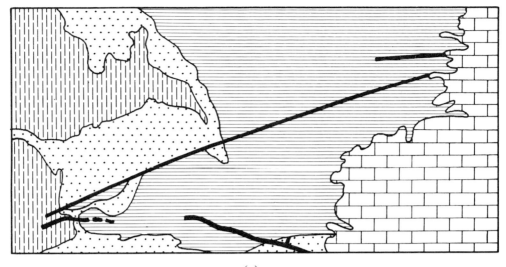

(a)

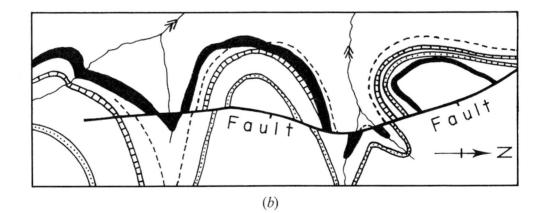

(b)

Figure 60 a. Dykes (shaded black) south of Durham. They intrude Carboniferous rocks, including Coal Measures, but end at the Magnesium Limestone (on the right); they are therefore post-Carboniferous but pre-Permian in age. The longest shown is 37 km from end to end.

b. Outcrop of the Whin Sill (shaded black) on the Carboniferous escarpment east of Carlisle. A few beds are shaded to show how the sill transgresses. The broken line indicates a coal seam. Scale about 1 : 40 000.

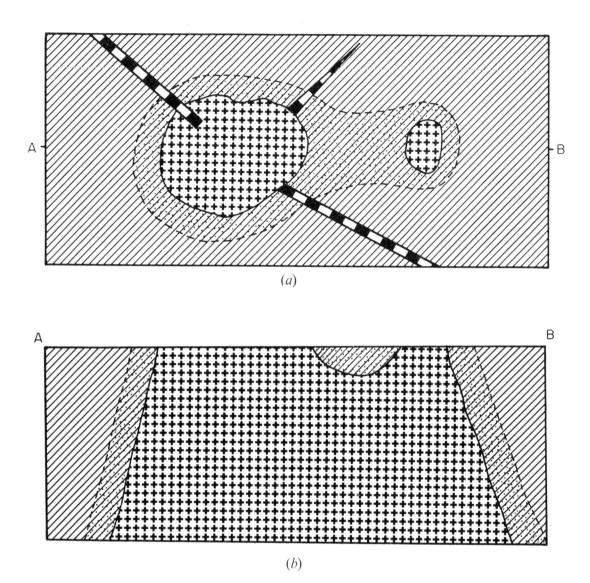

(a)

(b)

Figure 61

Many of the larger intrusions encountered in the British Isles are granitic. Figure 61 shows such a body in plan *a* and section *b*. At the surface there appear to be two small round bosses of granite (stocks are more irregular in outline), which are seen in section to form part of one mass underground with steeply dipping contacts. The *country rock* (in this case slate) surrounding them has been thermally metamorphosed in a zone indicated by stippling.

Note the vertical dykes of quartz-porphyry. They are usually shown as starting at the margin of the granite. Why should we suspect from the map that the bosses are connected underground?

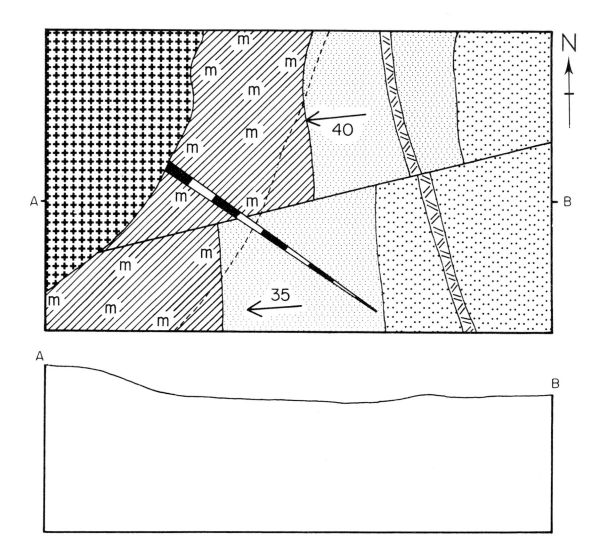

Figure 62

Exercise

Draw a section across the map from *A* to *B* on the profile provided, and write an account of the geological history of the area shown.

Is the granite and its accompanying dyke younger, or older, than the fault? The dolerite dyke is affected by the fault, but must be younger than the sedimentary rocks. List every event in order before you begin to write.

50

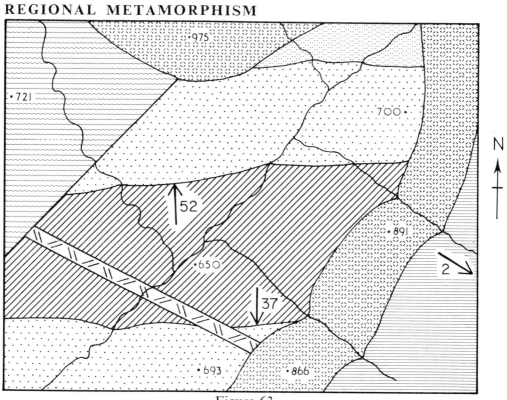

Figure 63

On the above map the unconformity and fault, both trending south-west to north-east, can be recognized. Between them is an anticline with east-west axis, slate being exposed in the core, grit and sandstone on the flanks. Across the fault to the north-west is an outcrop of gneiss. Basalt lies unconformably on both the gneiss and the folded rocks, and is followed with apparent conformity by shales. This unconformable series is almost horizontal and must have covered the whole area originally.

Gneiss is the product of intense regional metamorphism and must be older than any of the other rocks, underlying them where not exposed. Slate is formed from argillaceous rocks under conditions of low grade regional metamorphism which do not affect grits and sandstones appreciably. Since this second metamorphic episode did not alter the shales above the basalt, it took place before they were laid down.

Exercises

1. Write a geological history of the area shown in Figure 63.

Since gneiss is formed at depth, the unknown basal member of the folded series must rest on it unconformably. Are the dyke and the basalt associated in any way?

2. Draw a sketch section across the map (Fig. 63) from north to south through the 975 m spot-height.

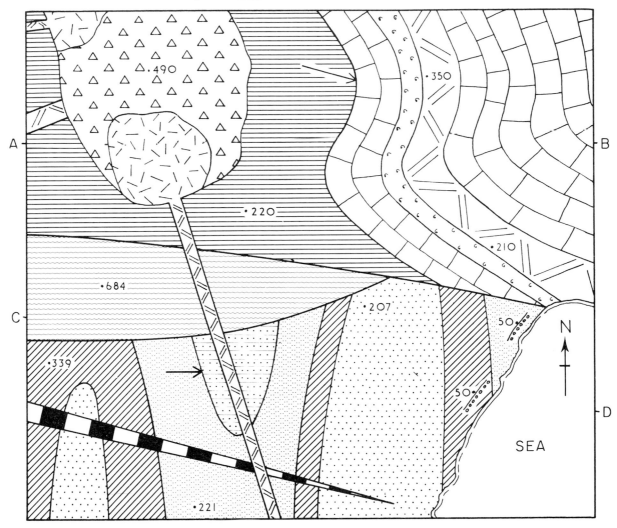

Figure 64

Scale 1 : 10 000

Exercise

Draw sections across the map from *A* to *B*, and from *C* to *D* to illustrate the geological structure, and give an account of the geological history of the area.

In drawing profiles across the map take into account the spot-heights shown, and the probable differential resistance of the rocks. The volcanic breccia or agglomerate represented by triangles is found in a volcanic neck; magma later found its way up through the agglomerate.

Note the deposits of gravel. What is the significance of their constant height above sea-level?

52

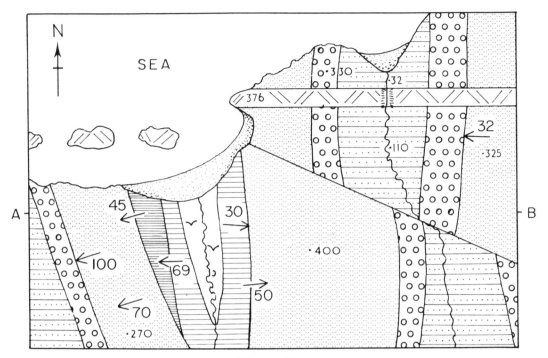

Figure 65

Scale 1 : 200 000

Exercise

On Figure 65, draw a section from *A* to *B* and describe the geology of the area, including the effect of structure on relief and land-forms.

How to describe the geology of the area shown

Formations present in order of age are: older shales; younger shales; sandstone, conglomerate, flags; dolerite (intrusive); alluvium, beach sand and pebbles. The shales outcrop in the westerly river valley beneath sandstone, and the relationship between the older and younger shales, though not clear, may be one of unconformity. A younger series of rocks rests unconformably on the shales, sandstone at the base being followed upwards by conglomerate and flags. This series is affected by N-S trending folds; an asymmetrical anticline with some overfolding in the west, in the core of which the shales are exposed, and an asymmetrical syncline in the east. The strike of the outcrops is therefore N-S. A dolerite dyke running E-W has been intruded in the north, and together with other structures has been displaced by a tear faulting trending WNW to ESE.

Two northward-flowing rivers have developed on shales and flags respectively, the more following along the anticline with a strip of alluvium along most of its length, meanders and an oxbow lake. The more easterly river following the synclinal axis deviates along the fault, and has cut a steep-sided valley through the dyke. Sandstone and conglomerate form the higher ground.

The sea has formed bays backed by beach sand and pebbles in the valleys at the mouths of the rivers. Parts of the dolerite sill resist erosion as islands.

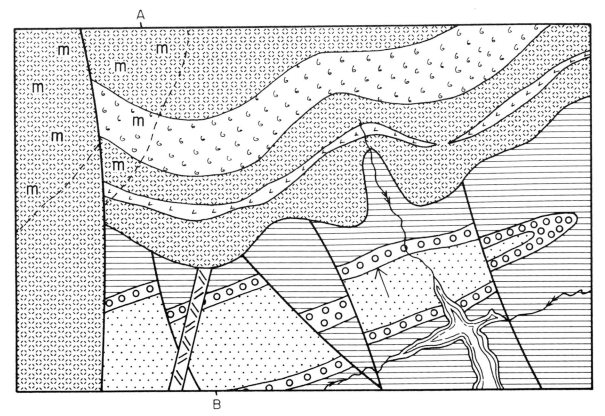

Figure 66

Exercise

Describe in chronological order the sequence of deposition, folding, faulting, igneous events and denudation shown on the map, Figure 66. Illustrate by drawing a sketch section from *A* to *B*, and mark the downthrow side of each fault.

This is a condensed sequence of the events portrayed on the map.

(i)   Deposition of sandstone, conglomerate and shale in that order as a series.

(ii)   Folding of the series into an anticline plunging down in the east.

(iii)   NNW-SSE trending faults developed.

(iv)   Uplift and erosion.

54

(v) Intrusion of the dolerite dyke, possibly as a feeder to the lava flows. Flows of basalt and sheets of ashes (now tuffs) covered the area.

(vi) A metamorphic aureole was formed in the north-east.

(vii) A N-S fault developed.

(viii) Uplift and erosion. Development of river system.

(ix) A relative rise in sea level forming a ria.

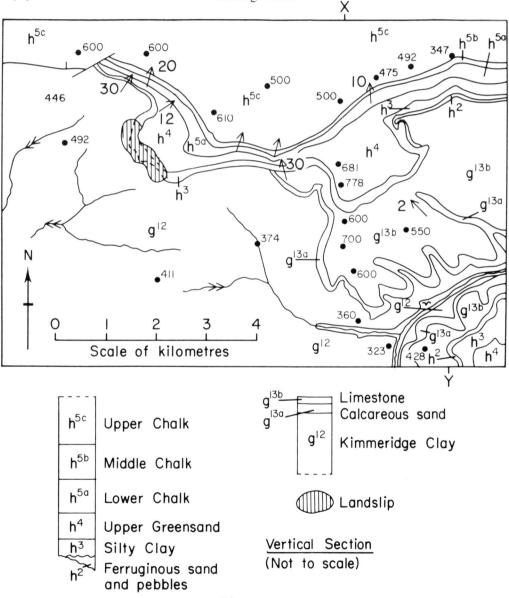

Figure 67

Figure 67 is taken, slightly modified, from part of the I.G.S. 1 : 50 000 Wincanton sheet 297 (England and Wales). The standard reference letters and indices used by the I.G.S. have been retained, so that colour or shading may be used on the vertical section and the map to identify formations readily.

In practice, even in limited areas dips are seldom uniform in size and direction, and formations may vary in thickness from place to place, so that it may be difficult to draw accurate sections. The sections shown below most 1 : 50 000 maps often incorporate additional information, derived from the logs of wells and boreholes, which is not available to the average user. In spite of these limitations, it is vital to be able to 'see' in three dimensions the underground structures shown on a map in two dimensions; and sketch-sections help us to do this.

On coloured 1 : 50 000 maps published by the I.G.S. contours are often extremely difficult to distinguish, and undue time should not be spent in ensuring accurate topography at the expense of geological structure: thus profiles are to some extent generalised.

Sketch-sections are not always drawn true to scale, but this should not be made an excuse for needless inaccuracy and untidiness. In Figure 67 spot-heights have been substituted for contour lines to avoid confusion with the geological boundaries.

Before attempting to draw any sections, the following should be noted:

(i) The silty clay, h^3, has been laid down on the bevelled edges of Jurassic rocks with h^2 as a local basal conglomeratic sandstone (see also Fig. 49). From information given in the vertical section or key, there must be unconformity between h^2 and h^3, but no evidence of this can be seen on the map.

(ii) The fault shown in the NW corner trending E-W is terminated by a small cross-fault (faults must end somewhere). This cross-fault throws down to the NNW, so terminating the outcrops of h^3 to h^{5b} abruptly. These formations continue beyond the fault beneath the Upper Chalk, h^{5c}.

(iii) Note how the Kimmeridge Clay, g^{12}, vees down the main valley in the SE, with tributary valleys on either side forming long, narrow minor vees.

(iv) The local, sudden widening of the outcrop of Upper Greensand, h^4, is caused by relief. This must be shown on the section.

(v) Slipped material conceals outcrops in one locality.

Exercise

Draw a sketch-section from X to Y, using spot-heights to draw the profile. Read the above notes carefully before attempting the exercise.

56

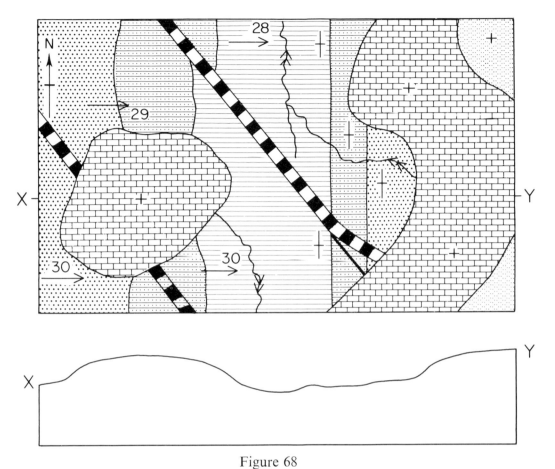

Figure 68

Exercise

Draw a section across the map from X to Y on the profile provided. Study the following way of describing the geological history of the area.

Geological history

(i)   A series of sedimentary rocks, with grit the lowest member seen, and followed conformably by flags and shales, was first laid down.

(ii)   These rocks were folded into a syncline trending N-S with a vertical easterly limb.

(iii)   A fault developed, trending NW-SE, downthrowing to the SW.

(iv)   Quartz porphyry dykes with a trend parallel to the fault were intruded, one following the fault for some distance.

(v)   Uplift and prolonged erosion followed.

(vi)   Subsidence below sea-level then took place.

(vii)   A newer series of rocks, limestone beneath followed by sandstone, were deposited.

(viii) Uplift and erosion again occurred, removing the newer series from the centre and west of the area, excepting one outlier. A drainage system developed.

57

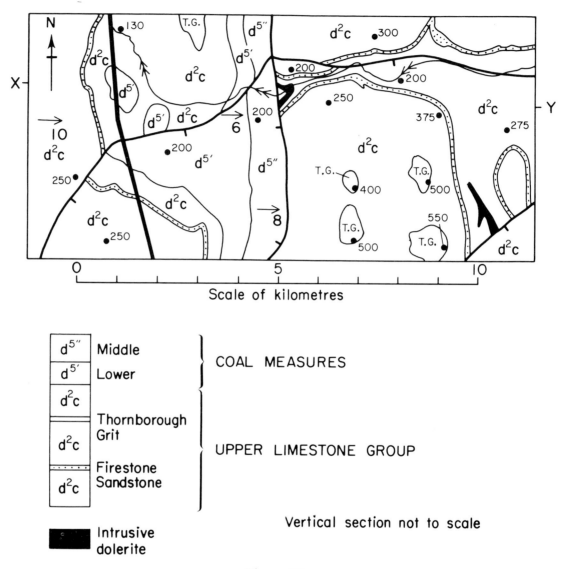

Figure 69

Figure 69 is based on part of the I.G.S. 1 : 63 360 map of Brampton, England and Wales Sheet 18, but has been simplified by a number of omissions. The key and map may be coloured or shaded if this helps the student. Symbols used are those employed by the I.G.S. on the original map.

The Coal Measures shown form part of the small Midgeholme coalfield, faulted down by the Stublick Fault to form an outlier. The Thornborough Grit caps several hill summits, also as outliers. Think carefully about the forms of intrusions in which dolerite is found before deciding that a dyke or sill is present. Then draw a sketch-section from X to Y to show the structure.

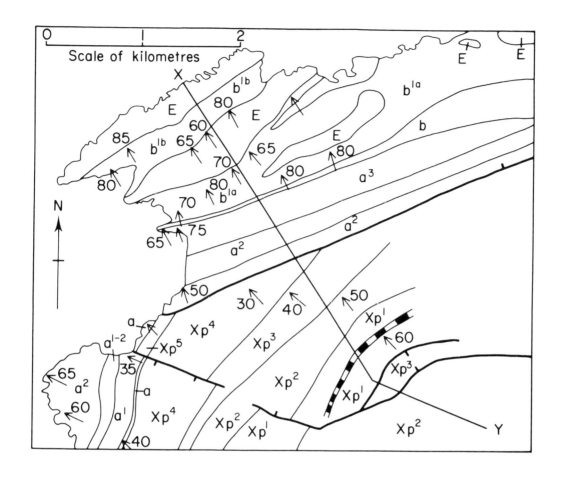

Figure 70

Figure 70 is a slight modification of part of the 1 : 25 000 I.G.S. sheet of St. David's. The line of section on both maps is the same, so that students may check their sections on completion. The Precambrian series of tuffs and lavas can be mapped as distinct groups in the west, but in the east exposures are so few because of the flat, drift-covered boggy nature of the surface, that no division is possible. Draw a section from X to Y on the profile provided.

Xp^1 to Xp^5 — Precambrian tuffs and lavas

a   to a^3 — Cambrian

b   to b^{1b} — Ordovician

E — gabbro, intruded as thick sills.

59

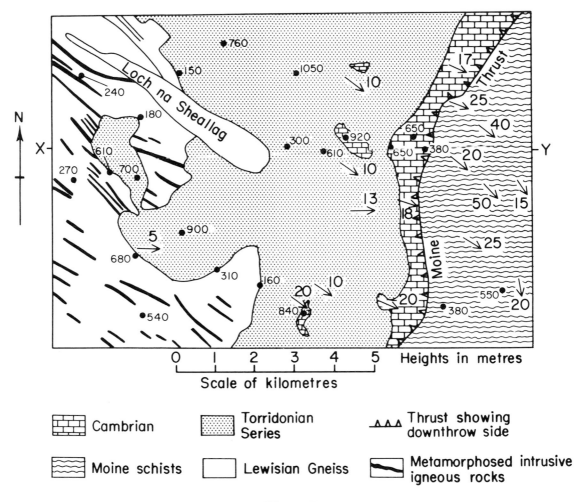

Figure 71

Figure 71 is a modified version of part of the I.G.S. sheet 92, (Inverbroom) Scotland. Lewisian gneiss exposed in the west, forms an irregular basement with Torridon Sandstone resting on its undulating surface. In turn Cambrian strata rest unconformably on the Torridonian, while the Moine Schist has been thrust westwards over the Cambrian.

Dips in both metamorphosed series are very variable in magnitude and direction, since much folding occurred during metamorphism. Variation of dip in Torridonian and Cambrian strata is not so marked.

Exercise

Draw a sketch section across the map from X to Y. The thrust shown (with a triangular symbol along it) is almost horizontal. Give a geological history of the area as a series of numbered statements. Comment on the trend of the metamorphosed intrusive igneous rocks.

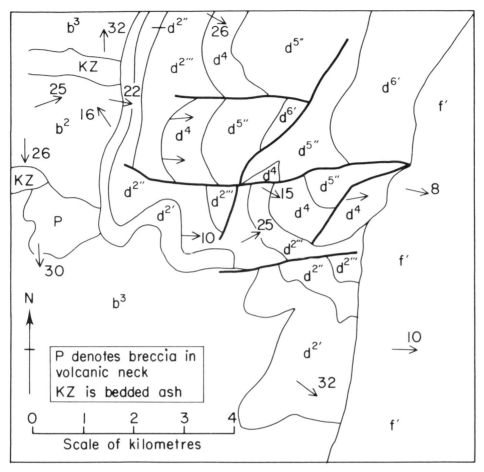

Figure 72

Figure 72 is a simplified adaption of part of the 1 : 50 000 I.G.S. map of Oswestry, Sheet 137 England and Wales.

Exercise

Study the map and answer the following questions:

1. The relationship between two rock formations may be *conformable, unconformable, concordant* or *discordant*. State which of these applies most correctly to
(a) the relationship between f′ and d²′ to d⁶′
(b) the relationship between P and b² to b³
(c) the relationship between KZ and d²′
(d) the relationship between d⁴ and d⁵.
2. Where possible mark the downthrow side on each fault.
3. The lettering and indices give the geological ages of the beds in sequence. What events happened between the deposition of KZ and d²′?
4. Draw a sketch section from X to Y to show the geological structure. The ground occupied by f′ is low-lying. The west is higher.

61

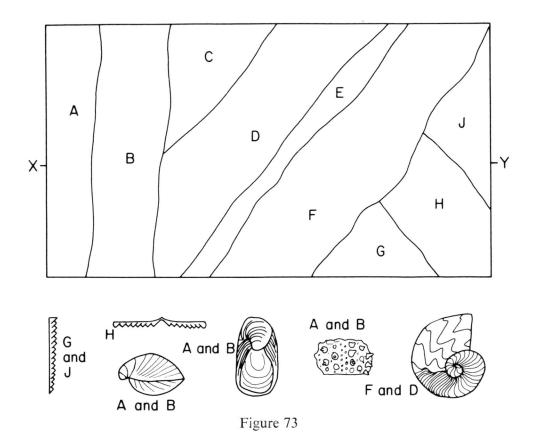

Figure 73

The formations shown on the map are lettered A to J. Fossils drawn below are labelled with the letter of the formation in which they are found.

Exercise

A specimen of rock from B is finely crystalline, easily scratched with steel, and effervesces with acid. Rock E is crystalline, of medium grain with dark green and white crystals. The shales near its margins are bleached and hardened, but concordant. Identify both rocks, and explain why the shales are altered. (The teacher may substitute hand specimens to illustrate the fossils and rocks described.) Give the probable geological ages of the fossils.

Insert arrows on each formation to show the direction of dip, and draw a sketch-section from X to Y to show the structure. No faults are present.

62

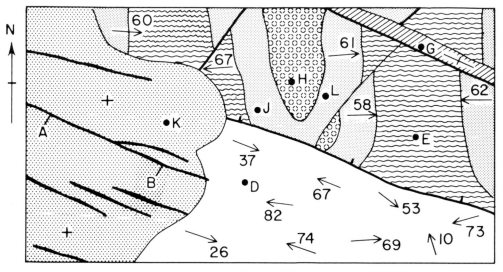

Figure 74

Labelled specimens provided for identification have been obtained from correspondingly labelled localities on the map. Symbols on the map do not represent rock types.

Exercise

1. Describe and identify specimens A and B. What does the thick line between A and B and elsewhere on the map represent?

2. Describe and identify specimens D, E, G, H and J.

3. Explain the shape and position of the rock body of which G is representative.

4. Identify as closely as you can the fossils K and L.

5. Give the geological history of the area as a series of numbered statements in sequence.

(Suitable specimens suggested for A and B include calcite, fluorite, galena, blende, barite, pyrite and quartz; for D, slate or schist; for E, H and J, sedimentary rocks; for G, a dyke-rock; for L, a Lower Palaeozoic fossil; for K, a Carboniferous fossil.)

Figure 75

Exercise

Measure the angle of dip of each limb of the fold shown in Figure 75, writing each angle against the appropriate limb. Draw the axial plane (axial plane trace) of the fold, as on page 13. Name the type of fold present, saying whether it is symmetrical or asymmetrical.

64

Figure 76

Exercise

Figure 76 shows a small fault with a zone of jumbled rock rather than a clear-cut fault plane. Compare the beds on either side of this fault zone, and state which is the downthrow side, left or right. Assuming that 1 cm on the photograph represents a distance of 1 m, measure and state the throw of the fault in metres.

Figure 77

Exercise

Figure 77 shows Hutton's classic unconformity at Siccar Point, Cockburnspath, Border. Measure and record the angle of apparent dip of the lower series of beds, measuring on the right-hand side; also of the upper, unconformable series. Now tilt the book sideways until the upper series is horizontal, and note how the apparent dip of the lower series has altered. Assuming that both series were deposited as horizontal beds, state the history of deposition, erosion and tilting shown by the rocks in the photograph. Say if the surface of unconformity is smooth or irregular.

Figure 78

Exercise

Figure 78 shows another unconformity at Portishead, Avon. Explain in what ways this differs from the unconformity shown in Figure 77. Points to consider in your answer are the types of rocks and bedding present, the dips of the older and younger series, and the plane of unconformity.

FIRST PUBLISHED IN 1966
METRIC EDITION 1973
SECOND EDITION 1977
THIRD IMPRESSION 1981

GEORGE ALLEN & UNWIN LTD
40 Museum Street, London WC1A 1LU

© George Allen & Unwin (Publishers) Ltd. 1966, 1973, 1977

ISBN 0 04 550024 X

Printed in Great Britain
by the Alden Press, Oxford